Motion and Matter

Full Option Science System
Developed at
The Lawrence Hall of Science,
University of California, Berkeley
Published and distributed by
Delta Education,
a member of the School Specialty Family

1487702
978-1-62571-312-4
Printing 3 — 5/2015
Quad/Graphics, Versailles, KY

Table of Contents

Magnetism and Gravity

Did you ever ride a skateboard, or watch someone else ride one? What makes the skateboard move? The idea is pretty straightforward.

You put one foot on the skateboard and give a good hard **push** with the other foot. The skateboard and rider start moving in the **direction** of the push. At least, that's what you expect will happen.

The skateboard will not move without a push. If you want to make the skateboard move without a rider, you can get down on the ground and give the board a push. It will move away without a rider. The secret to getting a skateboard to move is to give it a push.

A skateboard moves when a force is applied to it.

In science, pushes are known as **forces**. A skateboard will move only when a force is applied to it. There is a second kind of force that you probably know about, too. Another way to get a skateboard to move is to attach a string to it and **pull** it along behind you. A pull is another kind of force.

Pushes and pulls are forces. Forces change the **motion** of objects. Forces can start things moving. Force can make things move faster or slower. Forces can also change the direction something is moving. In fact, it takes force to make an object stop once it is moving.

You conducted investigations with **magnets** in class. You probably used a little force to push the magnet across your desk. The magnet moved until you stopped pushing it. You might have pushed the magnet over near a second magnet. When the magnet got close to the second one, one of two things probably happened. When the magnet got close, the second magnet might have jumped over to the first magnet and stuck to it. Or maybe when they got close, the second magnet started to move away. How can that happen?

Remember, to start an object moving you have to apply a force. But you didn't actually touch the second magnet. So where did the pushing force come from? It seems as though an invisible force pushed the second magnet.

That's exactly what happened. Let's find out more about magnets to help us explain this behavior. Every magnet has two different sides, or ends, called **poles**. One pole is called the north pole. The other is called the south pole. A simple bar magnet has its two poles on opposite ends. A horseshoe magnet has a pole on each end of the horseshoe. The doughnut magnets you worked with have poles on the two flat sides. Magnets always have a north pole and a south pole.

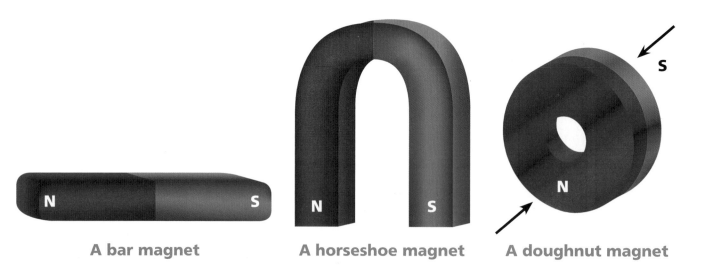

A bar magnet   **A horseshoe magnet**   **A doughnut magnet**

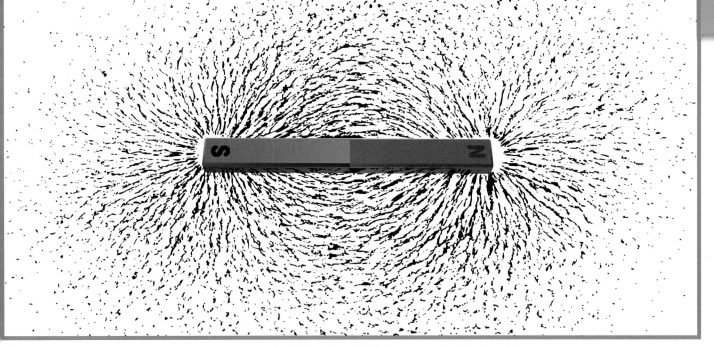

Magnets are surrounded by invisible magnetic fields.

Every magnet is surrounded by an invisible **magnetic field**. The field is made up of countless lines going out from the north pole in larger and larger loops and coming back to the magnet at the south pole. The poles of the doughnut magnets are on the round flat surfaces.

If the north poles of two magnets approach one another, the fields will push on each other. If the south poles of two magnets approach one another, the same thing happens. The same poles push on each other. The push is the **magnetic force**. When the magnetic force pushes magnets apart, the magnets **repel**.

What happens when the opposite poles of two magnets approach one another? If the south pole of one magnet approaches the north pole of a second magnet, the magnets pull on each other. The magnetic force will pull the magnets, and they will snap together. The magnets **attract**.

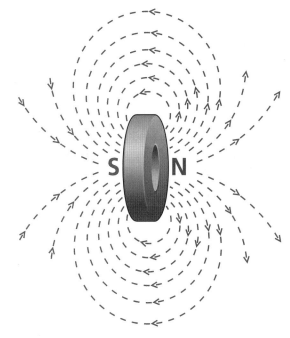

If you could see the magnetic field lines on a doughnut magnet, they might look like this.

If you hold a magnet a few centimeters above your desk and let go of it, what happens? It falls to the surface of your desk. Falling is a kind of motion. Falling objects usually move straight down. Objects fall because a force acts on them. Do you know the name of the falling force? It is **gravity**.

The force of gravity is the pull that makes everything move downward. Baseballs thrown up into the air come back to Earth, pulled by the force of gravity. Drops of rain that form in clouds fall to Earth because the force of gravity pulls them.

Gravity makes this ball come down.

Gravity pulls rain drops to Earth.

One of the other activities you did was to put magnets on a straw. You may have observed two different outcomes. Look at the illustrations on the right. These are the two outcomes that you might have observed.

Can we use what we know about magnetic force (**magnetism**) and the force of gravity to explain these two outcomes? Let's study the two **systems**. System A shows three magnets oriented so that they all attract one another. The magnetic force pulls them all together. The magnets slide on the post. Gravity pulls them down to the bottom of the post.

System B has more going on. The force of gravity is pulling the three magnets down toward the bottom of the post. But only the lowest magnet is resting at the bottom. The orientation of the three magnets causes the magnetic force to push them apart. They are repelling one another. Gravity is pulling the three magnets down toward Earth. The magnetic force is acting against gravity, pushing the magnets apart. You can see that the force is acting at a **distance** in system B. The magnets are pushing on each other, but they are not touching.

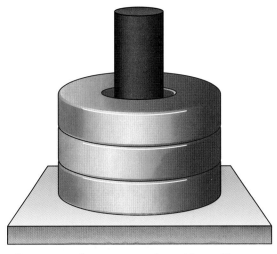

System A, magnets attracting

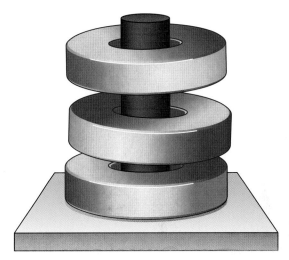

System B, magnets repelling

What makes these magnets appear to float?

7

What Scientists Do

Scientists are the men and women who answer questions about the natural world. Some scientists answer questions by making careful **observations** in the field. Other scientists answer questions by doing **experiments** in a laboratory. There are many scientists who do both. They make field observations, and they do experiments. But first, these scientists make observations. The observations lead to questions. Scientists use the **data** they collect from experiments and field observations to look for **patterns**. These patterns help them make general statements to explain what they observe in the natural world.

Investigating the **natural history** of the monarch butterfly is an example of observational science. Scientists go into the field to observe the behaviors of monarchs. Monarchs in North America migrate from northern areas to southern areas. During this migration, some of the butterflies are captured and fitted with tiny identification labels. When the labeled butterflies are later recaptured, their new location is recorded. Scientists follow thousands of monarchs over long periods of time. Then, the scientists can start to piece together the details of the butterfly natural history. Scientists have **evidence** that the life cycle of any individual butterfly is quite different from the migration cycle. The migration from the north to the south and back to the north takes much longer than any one butterfly can live.

An example of experimental science is your class investigation of questions about the **strength** of a magnetic field. The investigation started with some careful observations of how magnets and paper clips interact. First, you measured the distance that a paper clip jumps to one magnet. Then, you measured the distance that a paper clip jumps to three magnets. Based on these two observations, you **predicted** the jump distance between a paper clip and two magnets. You then tested your prediction and recorded your observations. This was your experiment.

Based on this experiment, a scientist might conclude that many magnets together create a stronger magnetic field. But a very observant scientist might notice something more. It is true that the magnetic field around two magnets is stronger than the magnetic field around one magnet. But it is not twice as strong. And the strength of the magnetic field around three magnets is even stronger. But the force is not three times as strong as the magnetic field around one magnet. The scientist might record the data like this.

Distance a paper clip jumps to magnets	
Number of magnets	Distance (cm)
1	2.5
2	3.0
3	3.5
4	4.0
5	4.3
6	4.4

These observations might lead the scientist to ask new questions. What is the effect of more magnets on the strength of the magnetic field? New questions require scientists to design new experiments. New experiments lead to new observations that will provide more information. More information helps scientists understand the interaction between magnetic fields.

Change of Motion

A wagon is a useful tool for moving a heavy load around. You might give your sister or brother a ride on the sidewalk. Or suppose you had a wagon sitting motionless with a load of pumpkins in it. To move the pumpkins, you will need to put the wagon into motion. How can you do that? You have two options. You can get behind the wagon and push. Or you can grab the handle in front and pull. The wagon will not move by itself. The wagon will move only if a force acts on it. Pushes and pulls are forces. Forces make things move.

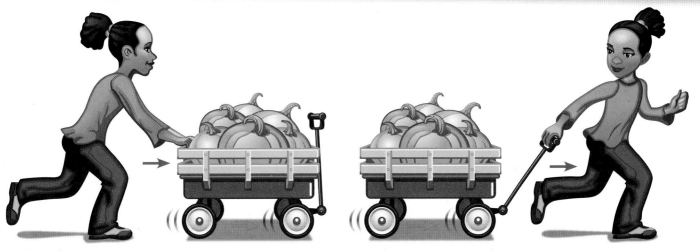

Pushes and pulls are forces (red arrows). Forces make things move (blue arrows).

If you use a force to get the wagon moving, it will keep rolling. But you don't want the moving wagon to crash into something. How can you stop it? It takes force to change the motion of a moving object.

Again, you can do one of two things. Look at the pictures below. You can get in front of the wagon and push to slow or stop its motion (a). Or you can get behind the wagon, grab it, and pull to slow its motion (b). To make a moving object stop, you need to push or pull in the opposite direction of the motion. To change the motion of an object, a force is needed.

a. Push to stop   **b. Pull to stop**

Each wagon was moving to the right (blue arrow). A force in the opposite direction (red arrow) can cause each wagon to stop.

If the rolling wagon of pumpkins is moving too slowly, can you make it move a little faster? You can if you use more force. If you get behind the rolling wagon and give it another push, the wagon will move faster. If you get in front of the wagon and give another pull on the handle, the wagon will move faster. A push or pull in the direction of the motion will make the wagon move faster. To change the motion of an object, a force is needed.

If the wagon starts moving too fast, use a push or a pull to slow it down. A force can cause a moving object to change its speed. If the wagon starts to turn to one side, how can you get it rolling straight again? Use a force. But this time, you need a push or pull to the side of the wagon to change its direction of motion. Any change of motion of an object, such as starting, stopping, change of speed, or change of direction, requires a force.

A force applied to the side of a wagon will change its direction.

Gravity

Think about a ball in one spot on a table. A gentle push on the ball will put it into motion. The ball will roll across the table. What will happen when the ball comes to the edge of the table? The ball will roll off the edge and fall to the ground. The ball's motion changes when it rolls off the edge of the table. It moves in a different direction and starts to move faster.

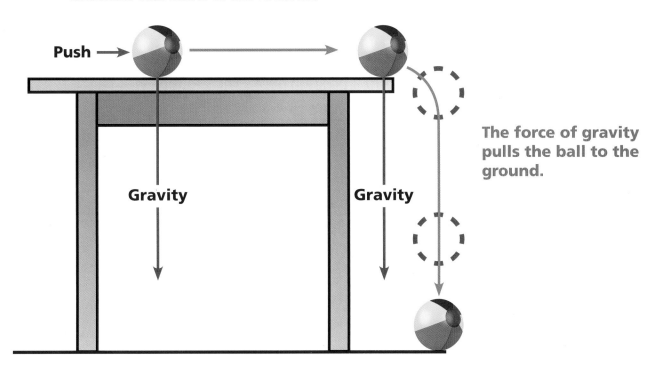

Push →

Gravity   **Gravity**

The force of gravity pulls the ball to the ground.

What causes this change of motion? That's right, force. What force makes the ball move toward the ground? The force that makes the ball fall to the ground is gravity. Gravity is a pulling force between two objects, and it draws them toward each other. As objects get bigger, the force of gravity between them gets stronger. Earth is a huge object, so it pulls strongly on all other objects. It is the force of gravity that pulls objects toward Earth's center.

But why doesn't the ball on the table move before you give it a push? Gravity is pulling on the ball, but it is not falling. The ball doesn't move because the table is pushing up on the ball. The table pushes up with a force **equal** to the force of gravity pulling down.

Balanced Forces

When two forces are exactly equal but push or pull in opposite directions, we say the forces are **balanced**. If you hold a ball up above your head, it will not fall to the ground as long as your arm muscles push up. Your muscles must push up with a force that is exactly equal and opposite to the force pulling the ball downward. After a short time, your arm muscles will tire. Your arm will no longer push up with a force equal to the force of gravity pulling the ball down.

But gravity never tires. Gravity always pulls down. Soon the forces keeping the ball in a position above the ground will no longer be balanced. The force of gravity will be stronger than your arm force and the ball will fall to the ground.

If you return the ball to the flat tabletop, it will again not move. Why doesn't the ball fall to the ground? The ball doesn't move because the forces acting on it are balanced. There are two forces. One force is the table. The table is pushing upward on the ball. The other force is gravity. Gravity is pulling the ball downward toward Earth's center. When two equal forces act on an object in opposite directions, the forces are balanced. When the forces acting on an object are balanced, the object's motion does not change.

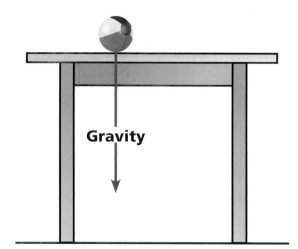

Gravity

The ball rests motionless on the table.

But what happens if you tip the table so it acts like a ramp? The ball starts to roll down the table. If the ball starts moving, a force must be acting on the ball. Tipping the table unbalances the forces. The forces are no longer equal and opposite. Gravity pulls the ball downhill toward Earth's center. The round ball rolls across the table, over the edge, and down to the ground.

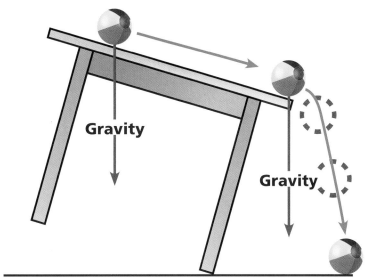

The force of gravity pulls the ball down the ramp to the ground.

Gravity

Gravity

Imagine you are sitting at the top of a slide. The moment the forces become unbalanced, gravity starts to pull you down.

Here's something to think about. What happens when you play baseball and hit the ball high into the air? Motion is involved, so there must be force involved. Let's analyze the activity.

The pitcher applies force to the ball with her arm. The ball moves in the direction of the batter. The batter applies force to the bat, which puts it in motion. If all goes well for the batter, the bat will make contact with the ball. Forces have both direction and strength. The direction and strength of the force applied by the bat sends the ball flying out into the field.

Can you think of other instances when forces of different sizes and direction are applied to an object? Think about soccer and bowling.

Patterns of Motion

You constructed **wheel-and-axle systems** in class. You used a **shaft** as an axle and attached a disk as a wheel on both ends. When you released the system on a ramp, the system rolled. When you construct a wheel-and-axle system using identical wheels, the system will roll straight down a ramp. If you replace one of the wheels with a smaller wheel and roll the system, something else happens. The wheel-and-axle system does not roll straight down the ramp. It rolls in a **curved** path.

What causes the system to roll in a curved path? The wheels in your system are attached securely to the ends of an axle. The wheels and axle **rotate**, or go around, together. Both wheels have to go around one complete time with each **rotation** of the axle. The wheels rotate exactly as often as the axle. If the wheels are the same size, both wheels go the same distance with each rotation of the axle. That's because it is the same distance around both wheels. But if the two wheels are different sizes, the distance around the larger wheel is farther than the distance around the smaller wheel.

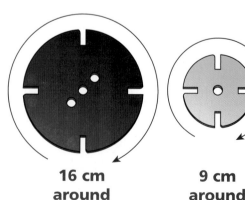

16 cm around   **9 cm around**

If you have a wheel-and-axle system with one large wheel and one small wheel, it is called an **uneven** wheel system. When an uneven wheel system rolls down a ramp, the larger wheel rolls farther with each rotation of the axle. This causes the system to follow a path that curves toward the smaller wheel.

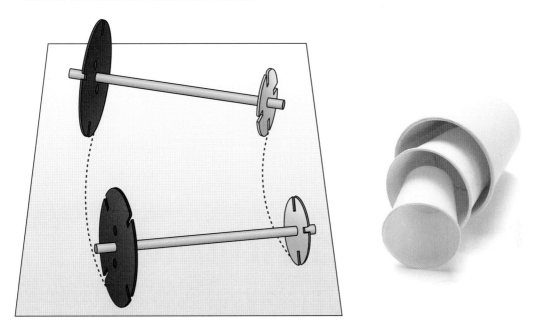

After you have observed many big and little wheel-and-axle systems roll down ramps, you will see a pattern. The pattern will allow you to predict the rolling path of other uneven wheel-and-axle systems that you have not yet seen roll.

You investigated cups as rolling systems. You probably figured out that a cup is just an uneven wheel-and-axle system. The open end of the cup is the larger wheel, and the base of the cup is the smaller wheel.

Thinking about Patterns of Motion

How do you think these objects might roll?

Football

Ice cream cone

Carrot

Flower pot

What Goes Around

The small spinner called a dreidel is a traditional Jewish toy. Children often play with dreidels during Hanukkah holiday celebrations. This toy is a little cube with a short shaft on top and a rounded point on the bottom. It has Hebrew letters written on the four flat sides.

The game is played on a smooth surface. Each player places a chocolate coin (or other small item) into the center of the table. When the first player takes the dreidel shaft between thumb and index finger, she applies a **rotational force** to the shaft to start the dreidel **spinning**. A balanced dreidel will spin for several seconds before it loses **energy**, slows, and falls over. When it stops moving, one of the sides will be facing up. The letter printed on the side

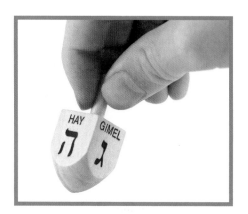

indicates what happens next. One letter means that the spinner gets all the items in the center of the table. Another letter indicates that the spinner gets half of the items. Each of the other letters means the spinner does something else.

The dreidel is just one of many traditional tops used by people around the world for entertainment. All tops are basically the same. A top has a central shaft around which it rotates. It has a **mass** that is **symmetrical**. The mass makes the top **stable**, so it can spin a long time. You designed a top by placing one or more disks on a plastic shaft. The mass of the disks made your top stable, so it could spin for several seconds.

The best way to apply force to this type of top is to spin the shaft between the palms of your two hands. You discovered that the position and size of the round disks changed the performance of your top. Which of these tops do you predict would produce the longest spinning action?

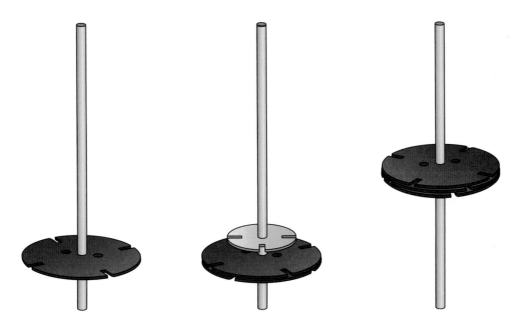

Here are some tops that work like the tops you designed in class. Notice where the round disk mass is located. Do these tops look like they will be stable spinners?

Another popular top design is like a dreidel. These tops all have a short shaft extending up from the disk mass. To put these tops in motion, you use your thumb and index finger. Some of them have colorful designs to create interesting visual effects when the top spins.

Many top designs use strings or a spring to apply rotational force to the top. To spin the string-driven tops, you wind the string around the shaft or the disk mass. Then you pull the string to apply force to the top to spin it. Tops that use string can rotate very fast.

Skittles and tops is a traditional board game. The skittle board is divided into several rooms. Doorways connect the rooms. The doorways are just big enough for the spinning top to pass through. In the rooms are ten little pins. Each pin is positioned on a spot that has been assigned a number. One of the rooms is the starting room. When a player pulls the string wound around the top, it starts spinning. The object is for the top to wander from room to room, knocking down the pins as it goes. When the top finally falls over, the player adds up the numbers of fallen pins. The highest score is the winner. The award for winning is figured at the start of the game by the players.

A steam-powered locomotive

What Engineers Do

Engineers solve problems. They solve problems about how to make something or how to fix things. Do you know an **engineer**? What does he or she do?

Engineers Operate Systems

You have probably heard of the engineer that drives the train. What kinds of problems could an engineer solve in the driver's seat of the train? We have to remember that old-time trains were powered by steam. The train had a big wood or coal fire burning under a boiler. The boiler turned water into steam. The steam moved into big cylinders where the pressure of the steam pushed a piston back and forth. The piston provided the force to turn the wheels.

The steam locomotive system was very complex. The fire had to be maintained at a steady temperature. The steam had to be delivered to the cylinders at just the right pressure. All the mechanical parts and connections in the system had to operate precisely for the train to keep chugging along. A moving train presented many opportunities for problems. It was essential to have an engineer there to solve problems while the train was moving.

Modern trains have huge diesel engines. The diesel engines do not, however, turn the wheels. The diesel engines turn generators. The electricity generated on the train powers electric motors connected to the wheels. The electric motors make the train go. The train you see rumbling down the tracks is actually a giant electric train.

A diesel locomotive is still a very complex system. It requires a skilled problem solver to keep the train moving properly. The engineer no longer attends to fire temperature or steam pressure. Instead the train engineer monitors the operation of the diesel engine and the electric power coming from the generators. The engineer monitors the signal systems that keep the train safe. In addition, the engineer watches other parts of the train system, such as stations, railroad crossings, and track condition.

A diesel-powered locomotive

Engineers Design and Improve Systems

Another type of engineer is someone who helps to design or improve systems. Engineers work on all types of systems, including transportation systems. There were engineers who designed the first diesel locomotive. Like other engineers, they set out to solve a problem. The traditional steam locomotive was dangerous and dirty to operate. Burning coal produced a lot of dirty, dark smoke. The boiler might explode if it got overheated. The rail industry needed a new way to power the locomotives that pulled the long trains full of passengers and freight.

The famous inventor and engineer Thomas Edison (1847–1931) led a team of engineers at General Electric. This team built their first electric locomotive **prototype** in 1895. There were many advances in engineering during the next 40 years. The Burlington and Union Pacific Railroads began using diesel "streamliners" to transport passengers in 1934. Diesel-electric railroad locomotion soon became widespread in the United States.

Today, engineers are designing new technologies to improve train transportation. These **technologies** are solving problems such as traction on the rails, braking time, and energy efficiency. Some trains are powered only by electricity (no diesel engines). Some designs even make levitating trains glide on air. Some engineers are designing "moving platforms" that dock with high-speed trains.

Thomas Edison was a famous engineer and inventor.

A maglev train uses magnetic fields to lift and drive the train.

24

These engineers are designing a robot.

Engineering Design Practices

When faced with a problem, engineers first define the problem carefully. They decide what might make a good design to solve a problem. The characteristics of a good design are the **criteria** for a **solution**. Here is what the criteria for a solution to the locomotive problem might include.

- Powerful enough to pull a long line of train cars
- Running on an easy-to-use fuel that doesn't cost too much
- Operating on rail systems in place now

The engineers must also consider limits on the solution. The limits are the **constraints** placed on the solution design. Here is what the constraints placed on the locomotive problem might include.

- Made of materials that are easy to get
- Not too expensive to manufacture
- Designed and produced in a short amount of time
- Safe for operators, passengers, and the environment

The first step to solve an engineering problem is for a team of engineers to understand the criteria and constraints. The criteria and constraints frame a solution to a problem. Next the engineers spend time developing a plan for the design of a solution. Once they agree on a plan, they assemble the materials and tools. Then they build a prototype. Once the prototype is built, they test it to see if it meets the criteria of a good solution. If the solution performs well, that is great. If the solution fails to measure up to the criteria, the team goes back to the original plan. They revise the plan to correct the parts of the design that didn't work. At the same time, they make sure that they honored the constraints. Is the solution cost-effective? Are all of the materials easily available? Is it safe and easy to operate?

Think about the engineer who designed the first parachute. The problem was how to get objects and people from airplanes gently but quickly to the ground. He had an idea and developed a plan. He made a prototype parachute. He tested it by dropping it from various heights with different masses attached.

An engineer designed parachutes.

Engineers thoroughly test their designs before putting them into production.

Think about the criteria for a successful solution to this problem of delivering objects and people from airplanes. The first test with bags of sand or potatoes was successful, and the design was judged to be good. Then someone had to put the new design to the ultimate test. Someone had to jump out of an airplane 1,000 **meters (m)** or more up in the air, using the parachute. That's one instance where the first user has to be pretty sure all the mistakes have been corrected.

Elements of the Engineering Design Process

Let's review the elements of the engineering design process.

1. Understand the problem thoroughly.

2. Carefully define the criteria and constraints placed on a solution.

3. Devise a plan for a solution.

4. Build the planned solution.

5. Test the solution and evaluate its performance.

6. Return to the planning phase and revise the plan, based on data from the test.

7. Repeat Steps 4–6 until the solution satisfies the criteria and constraints.

8. Obtain a patent and go into production.

How Many Kinds of Engineers Are There?

There are many different problems to solve. Every kind of problem is the specialty of a different kind of engineer. Here are some different kinds of engineers and what they do.

Architectural Engineers

Architectural engineers design buildings and structures such as homes, hospitals, skyscrapers, warehouses, towers, and stadiums.

Architectural engineers use blueprints to help them design a project.

Aerospace Engineers

Aerospace engineers design airplanes, rockets, satellites, space stations, and space shuttles.

The International Space Station that orbits Earth was designed by aerospace engineers.

Electrical Engineers

Electrical engineers design electric circuits that perform all kinds of electronic wonders. Their designs create cell phones, digital cameras, computers, televisions, and lighting systems.

The electronic components of a computer are designed and built by electrical engineers.

Chemical Engineers

Chemical engineers design new materials for use by people in many different ways. They design fabrics, medicines, lubricants, and fuels.

Chemical engineers design and test new products in a laboratory.

Biomedical Engineers

Biomedical engineers design new replacement parts, such as prosthetics, for people. They design instruments for monitoring a person's health, and new lifesaving devices. They also design processes or therapies to help sick people get healthy.

Designing artificial limbs, or prostheses, is one way biomedical engineers help to improve a person's health and wellness.

Mechanical Engineers

Mechanical engineers design machinery such as locomotives, cars, and motorcycles. They design smaller machines like printing presses, dental drills, electric toothbrushes, chain saws, braking systems for bicycles, and just about every other designed system that has moving parts.

Large and complex machines are developed by mechanical engineers.

Computer Engineers

Computer engineers design computing systems, including both the hardware and software components in computers and tablets. They design controller systems for cars, GPS systems for navigation, and cellular telephone systems.

Computer engineers design computer hardware and software.

Traffic Engineers

Traffic engineers design systems for the efficient movement of vehicles and people. They design roadways and rail systems, and smaller escalators to move humans through airport terminals and sports stadiums. These engineers manage aircraft from take off to landing, and design and manage shipping through river systems.

Air traffic control is managed by traffic engineers to make sure aircraft take off and land safely.

Acoustical Engineers

Acoustical engineers design the interior space and surface materials of theaters, concert halls, and recording studios.

Acoustical engineers observe the sound and vibration in a recording studio.

Nautical Engineers

Nautical engineers focus on the design of water vehicles, such as boats, military ships, cruise ships, and submarines. They also design docking facilities and safety equipment. They develop standards for use of equipment at sea.

Nautical engineers design and build large ships.

Civil Engineers

Civil engineers design the support systems of people living in communities. They design water-delivery systems and sewage systems. They design highway systems, bridges, dams, levees, canals, and tunnels. They also design public utilities systems that provide electricity and natural gas.

Civil engineers often survey construction sites.

Science Practices

1. **Asking questions.** Scientists ask questions to guide their investigations. This helps them learn more about how the world works.

2. **Developing and using models.** Scientists develop models to represent how things work and to test their explanations.

3. **Planning and carrying out investigations.** Scientists plan and conduct investigations in the field and in laboratories. Their goal is to collect data that test their explanations.

4. **Analyzing and interpreting data.** Patterns and trends in data are not always obvious. Scientists make tables and graphs. They use statistical analysis to look for patterns.

5. **Using mathematics and computational thinking.** Scientists measure physical properties. They use computation and math to analyze data. They use mathematics to construct simulations, solve equations, and represent different variables.

6. **Constructing explanations.** Scientists construct explanations based on observations and data. An explanation becomes an accepted theory when there are many pieces of evidence to support it.

7. **Engaging in argument from evidence.** Scientists use argumentation to listen to, compare, and evaluate all possible explanations. Then they decide which best explains natural phenomena.

8. **Obtaining, evaluating, and communicating information.** Scientists must be able to communicate clearly. They must evaluate others' ideas. They must convince others to agree with their theories.

Scientists ask questions and communicate information. Are you a scientist?

Engineering Practices

1. **Defining problems.** Engineers ask questions to make sure they understand problems they are trying to solve. They need to understand the constraints that are placed on their designs.

2. **Developing and using models.** Engineers develop and use models to represent systems they are designing. Then they test their models before building the actual object or structure.

3. **Planning and carrying out investigations.** Engineers plan and conduct investigations. They need to make sure that their designed systems are durable, effective, and efficient.

4. **Analyzing and interpreting data.** Engineers collect and analyze data when they test their designs. They compare different solutions. They use the data to make sure that they match the given criteria and constraints.

5. **Using mathematics and computational thinking.** Engineers measure physical properties. They use computation and math to analyze data. They use mathematics to construct simulations, solve equations, and represent different variables.

6. **Designing solutions.** Engineers find solutions. They propose solutions based on desired function, cost, safety, how good it looks, and meeting legal requirements.

7. **Engaging in argument from evidence.** Engineers use argumentation to listen to, compare, and evaluate all possible ideas and methods to solve a problem.

8. **Obtaining, evaluating, and communicating information.** Engineers must be able to communicate clearly. They must evaluate other's ideas. They must convince others of the merits of their designs.

Engineers use models.

Soap Box Derby

Have you heard about the Soap Box Derby? It is an organized race competition for young people. Boys and girls between the ages of 10 and 17 can compete for fun and prizes. The race cars are homemade. They must all use the same power supply to put them in motion, so the race is very fair. The derby cars have no motors. Soap Box Derby cars are gravity racers. The cars race downhill, and the force of gravity pulls the cars down the track.

Why is the race called the Soap Box Derby? In the beginning, derby racers were made from anything that was available. Often the main body of the racer was made from a packing crate or shipping box. Wooden soap boxes and orange crates were popular bodies for a racer. The designers had to scrounge around to find wheels and axles to fit the racer.

Starting in the 1930s, the derby racers began to set standards. Today, races are conducted in four categories or divisions. The Stock Division is for boys and girls 7 to 13 years old. Racers in this division have to be built from a kit of standard components.

The Super Stock Division is for boys and girls 10 to 17 years old. Racers in this division use the same basic components as the Stock Division. But the cars can be modified to show a little individuality.

The Masters Division is for boys and girls 10 to 17 years old, also. Racers in this division use the standard wheels. However, the racers are allowed to express their creativity and design skills.

The fourth category is the Ultimate Speed Challenge. In this race, the goal is not to beat the other racers head-to-head. Rather, the racers need to beat the clock. The prize is awarded to the fastest racer. The competition is to see whose car can roll down the 301-meter (m) track in the shortest time.

Three racers in the 2010 Ultimate Speed Challenge

The first Ultimate Speed Challenge was held in 2004. Race car designers could modify their cars with custom wheels and tires. They could add wheel fenders and use streamlined body design. A car designed and built by the Pearson family was driven to victory by Alicia Kimball. She made the run down the official track in Akron, Ohio, in 27.190 seconds. The table shows the names of the winners since 2004 and their winning times.

Study the table. What pattern do you see?

Year	Winner/driver	Winning time (seconds)
2004	Alicia Kimball	27.190
2005	Niki Henry	26.953
2006	Jenny Rodway	26.934
2007	Lynelle McClellan	27.160
2008	Krista Osborne	27.009
2009	Jamie Berndt	26.924
2010	Sheri Lazowski	26.844
2011	Sheri Lazowski	26.585
2012	Laura Overmyer	26.655
2013	Anne Taylor	26.929

How do you think the race organizers set up the races to make sure they identify the fastest racer? You know from your experiment with rolling carts in class that position on the ramp is very important. The official track in Akron, Ohio is 301 m from start to finish line. The track has a carefully designed starting mechanism that holds the race car on the slope. When the starter pushes the switch to release the car, it also starts an electronic timer. The car is pulled down the slope by gravity all the way to the finish line. When the car crosses the finish line, it signals the timing system. The timing system records the length of time to the nearest 1,000th of a second.

Once the time is recorded, the next car is ready to race. The timer is reset. Race car number 2 goes rolling down the track. The cars all start from the same exact location on the ramp. They are released in exactly the same way. The timer starts at exactly the moment the car is released. The factors of release position, release method, elapsed time, and track surface conditions are controlled to be exactly the same for every competing car.

So what makes one car get down the hill a little faster than all the others? The car that wins is the one designed to minimize **friction**. Friction is the force that works against gravity to slow the car down. The other important factor is the skill of the driver. A driver who guides the car the straightest will go a little faster than a driver who has to steer back and forth to stay in the center of the track.

Soap Box Derby racing is a good example of how science and engineering work together. The scientist understands how friction and gravity work together as the car rolls down the slope. And the engineer understands that success means solving the problem of reducing the friction that slows the car as it rolls down the slope.

The Metric System

The metric system is an easy system of measurement to use. Can you count by tens? Can you multiply and divide by tens? Then you can use the metric system.

Measurement systems based on multiples of ten were proposed many times in history. In 1793, people in France created the metric system. The French based this system on a unit they called the meter (m). *Meter* comes from the Greek word *metron*, which means measure.

How did the French set the size of the meter? They made the meter one ten-millionth of the distance from the North Pole to the equator. They wanted the meter to be based on a unit that would never change. Today the meter is based on how far light travels in a fraction of a second.

The meter was used to create other metric units. The unit of mass is the **gram (g)**. All **matter** has mass. The unit of **volume** is the **liter (L)**. All matter takes up space and has volume.

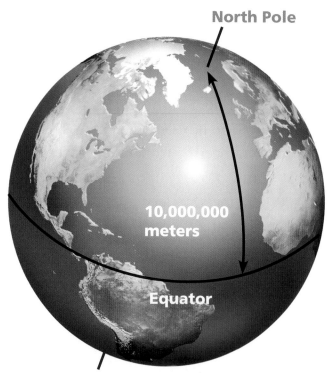

North Pole

10,000,000 meters

Equator

Metric Prefixes

All metric units are based on the meter. The prefix can help you tell how big a metric unit is. The prefix is the part of the word that comes first.

millimeter	=	0.001 meter (one thousandth)
centimeter	=	0.01 meter (one hundredth)
decimeter	=	0.1 meter (one tenth)
meter	=	1.0 meter
dekameter	=	10.0 meters
hectometer	=	100.0 meters
kilometer	=	1,000.0 meters

The metric system slowly caught on around the world. Seventeen countries signed the Treaty of the Meter in 1875. This treaty created the International Bureau of Weights and Measures. The bureau adopted the metric system as the worldwide standard of measurement. Today the metric system is the standard everywhere in the world.

But the metric system is not the standard in the United States. It is the only major country in the world that does not use the metric system as its official measuring system. But even in the United States, the metric system is used in many areas. It is used in most scientific fields. It is used in many sports and recreational activities. And one day, the metric system might be used for everyday measurement in your home.

Scientists use the metric system every day.

Length, Mass, and Volume

The meter is used to define the basic units of mass and volume in the metric system. Here's how.

Mass The basic unit of mass in the metric system is the gram.

One cubic centimeter of water has a mass of 1 gram.

Volume The basic unit of volume in the metric system is the liter.

A 10-centimeter cube has a volume of 1 liter.

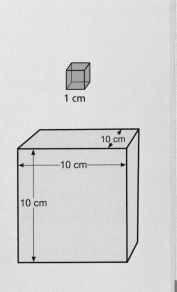

How Engineers and Scientists Work Together

Scientists gather data to answer questions about the natural and designed world. Engineers design and test solutions to problems. Engineers base their work on the work of scientists. And scientists need engineers to develop tools and methods to answer questions.

Here's an example of students working as both scientists and engineers. Some students were designing rolling carts and sending them down ramps. Team Thunder's cart went down the ramp and then rolled 2 meters (m) across the floor. Team Glider made a cart that was much like the Thunder cart. However, the Glider cart rolled only 1.6 m after it went down the ramp.

The Glider team wanted to know why their cart didn't roll as far as the Thunder cart. They asked the Thunder team if they would roll the Glider cart down the Thunder ramp. When the Glider cart was released on the Thunder ramp, it rolled 2 m across the floor. The Glider team wanted to know how to get their cart to roll 2 m on their own ramp. They decided to conduct some experiments.

The Glider team made a list of the things that might affect the rolling distance. They thought the ramp is what made the difference. Here are the factors they identified.

- The starting position on the ramp
- The slope (steepness) of the ramp

The team came up with this design for an experiment.

1. Repeat the original cart run. Start the cart from exactly the same position used before [10 centimeters (cm)]. Measure the roll distance.
2. Move the starting position 5 cm higher on the ramp (15 cm). Release the cart. Measure the roll distance.
3. Move the starting position another 5 cm higher on the ramp (20 cm). Release the cart. Measure the roll distance.

The team conducted the experiment. They found that the release position affected the roll distance. The higher the release position, the farther the cart rolled across the floor. These are the data they recorded.

Release position (cm)	Distance rolled (m)
10	1.5
15	1.9
20	2.2

The team analyzed their data to look for patterns. The starting-position data provided evidence to support their claim. Their claim was that the higher a cart starts on a ramp, the farther it rolls across the floor.

The team of students went on to design a second experiment to investigate the factor of ramp slope. How do you think they designed their experiment? What do you think they discovered?

Magnets at Work

Design solutions to many different problems use the magnetic fields that surround magnets. You have probably seen magnetic fields at work. Have you looked closely at the inside of a refrigerator door? The door on a refrigerator does not have a mechanical latch to keep it closed. Behind that rubber gasket on the door are magnets. Other magnets are hidden behind the edge of the refrigerator cabinet. That's the place where the edge of the door makes contact. The magnets are positioned so the poles attract. The magnets on the door are attracted to the poles of the magnets in the edge of the cabinet. The force of attraction between the magnets is strong enough to hold the door closed.

Magnetic closures keep kitchen cabinets closed.

Magnetic closures are also used to hold cabinet doors closed. A strong magnet is fixed on the inside of the cabinet. A little steel plate is fixed on the inside surface of the cabinet door. The attraction between the magnetic field of the magnet on the cabinet and the steel plate on the door holds the door closed. But the strength of the magnetic force is easily overcome by a pull on the cabinet handle.

Magnetic closures are used on small boxes, jewelry, wallets and purses, briefcases, and birthday cards. Can you find magnetic closures at your school or in your home?

Magnetic closures are used on many common items.

Magnets that you have probably never seen are at work on farms. A strong, smooth magnet about the size of an AA battery is put into the stomach of young cows. Why? While they graze in the field, cows often eat bits of dangerous steel trash. They might eat bits of baling wire, staples, or nails. If this trash got into the cow's intestines, it could cause a serious injury. The magnet attracts the bits of metal, holding them safely in the cow's stomach. The magnet stays inside the cow for its entire life, doing its work completely out of sight. The cow stays healthy.

These tiny magnets are put inside a cow's stomach to keep it healthy.

Observing a magnetic stirrer in the lab

Chemists often mix chemicals in their labs. They use a mechanical stirrer to mix them thoroughly. One kind of stirrer uses two magnets. One magnet, covered in plastic, is placed in a beaker with the **mixture** of chemicals. The beaker is then placed on a platform. Under that platform is a second magnet attached to an electric motor. When the platform magnet is turned on, the motor rotates the magnet. The magnetic field of the platform magnet interacts with the magnet in the beaker. The magnet in the mixture of chemicals rotates at the same speed as the magnet hidden inside the platform, stirring the contents of the beaker.

These are a few ways that magnetic fields surrounding magnets are used to solve problems.

Thinking about Magnets at Work

Can you find other ways magnetic fields are used to solve problems?

Mixtures

If you visit a lake or beach, you might see something like this at the water's edge. What's there? A mixture. A mixture is two or more materials together. This beach is a mixture of sand and gravel. A handful of this mixture contains bits of rock of many different sizes.

If you wanted to **separate** the gravel from the sand, how could you do that? You could pick out all the pieces of gravel one by one. But there is a faster way. You could use a **screen**. A screen has holes small enough for sand to fall through. Pieces of gravel, however, are too large to pass through. They stay on top of the screen. Screens are useful tools for separating mixtures based on the **property** of size.

A screen can separate sand and gravel.

Imagine opening a kitchen drawer to get a rubber band. Oops, the rubber bands spilled. So did a box of toothpicks and a box of paper clips. The drawer contains an accidental mixture of rubber bands, toothpicks, and paper clips. How can you separate the mixture?

You could use the property of shape. You could pick out each piece one at a time. It might take 10 minutes to separate the mixture.

Paper clips are made of steel. Steel has a useful property. Steel sticks to magnets. If you have a magnet, you can separate the steel paper clips from the mixture in a few seconds. Magnetism is a property that can help separate mixtures.

What about the toothpicks and rubber bands? Wood **floats** in water. Rubber **sinks** in water. The properties of floating and sinking can be used to separate the wood toothpicks and rubber bands in seconds. Drop the mixture into a cup of water. Then scoop up the toothpicks from the surface of the water. Pour the water and rubber bands through a screen. The water will pass through the screen, but the rubber bands won't. The job is done.

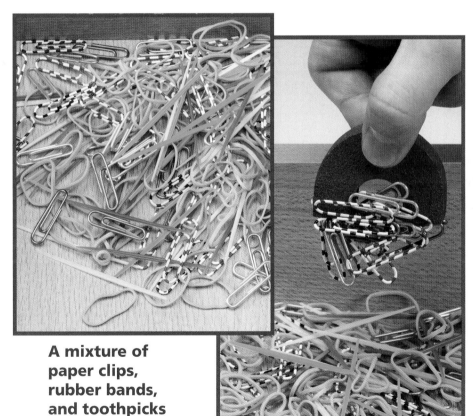

A mixture of paper clips, rubber bands, and toothpicks

Separating steel paper clips with a magnet

Separating toothpicks and rubber bands in water

Solids and Liquids

Mixtures of **solids** and **liquids** are interesting. Several things can happen. When sand and water are mixed, the sand sinks to the bottom of the container. If you stir the mixture, things move around, but that's about it.

When you mix **chalk** and water, the chalk makes the mixture **cloudy** white. After a while, the chalk settles to the bottom.

When you mix **salt** and water, the salt disappears, and the mixture is **transparent** and colorless.

Sand, chalk, and salt all make mixtures with water. After stirring, you can still see the sand and chalk, but the salt has disappeared. Salt is different in some way.

A mixture of salt and water forms a **solution**. A solution is a special kind of mixture. When solid salt and liquid water are mixed, the solid disappears into the liquid. The solution is transparent.

When the solid salt disappears in the water, it is *not* gone. It has **dissolved**. When a solid dissolves, it breaks into pieces so tiny that they are invisible. When salt dissolves in water, it makes a saltwater solution.

Mixing sand and water

Mixing chalk and water

Mixing salt and water

Sand mixture after 5 minutes

Chalk mixture after 5 minutes

Salt mixture after 5 minutes

Conservation of Matter

There's one more thing to think about when you make a mixture. All matter has mass. Anything that has mass is matter. If you have 50 milliliters (mL) of water in a cup and add 30 grams (g) of sand to the cup of water, what will the mass of the mixture be? The mass of the 50 mL of water is 50 g, so the mass of the mixture will be 50 g (water) + 30 g (sand) = 80 g (mixture). That seems pretty easy to understand.

50 g water   +   30 g sand   =   80 g mixture

But the mixture of salt and water is a little trickier to think about. When you mix 30 g of salt with 50 g of water, what do you think the mass of the mixture will be? The salt disappears in the water, so what happens to its mass? Did you conduct this investigation? The mass of the clear solution is 80 g. That is the sum of the mass of the water (50 g) and the mass of the salt (30 g). The mass of a substance like salt does not change when it dissolves. Even if you can't see it, the salt is still there, and its mass has not changed. In fact, mass never goes away. Mass is **conserved**, therefore matter is conserved. That means matter can change shape, state, or location, but it can never be lost or destroyed.

50 g water   +   30 g salt   =   80 g salt mixture

Matter is never destroyed, but it can change. Wood (matter) changes to ash when it burns.

Sometimes it is hard to understand how matter is conserved. For instance, when you have a campfire, a large mass of wood burns and all that is left at the end of the evening is a small pile of ash.

If matter is conserved, where did the mass of the wood go? The fire produced several things. It produced smoke, light, and heat. Light and heat are energy. Energy is not matter. Smoke is **gas** and tiny particles of soot. Gas and soot are matter. That's where the wood went. The fire changed most of the mass of the wood into gas and tiny particles. The particles drifted off into the air. Gases and tiny particles have mass.

If you could capture all the smoke and dust coming up from the fire, and gather up all the ashes, what would you find? You would find that the mass of the gas and ash would add up to the mass of the wood you put on the fire earlier. Conservation of matter is just one of the great truths of nature. Matter can never be destroyed, but it can be changed.

As you continue your investigations of mixtures, you might find an instance where your observations suggest that matter is not conserved. But matter is always conserved. You will have to do some deep thinking to explain why your evidence suggests that matter is conserved.

Reactions

Vinegar and baking soda are two materials you have worked with in class. Vinegar and baking soda have properties that help you identify them. Vinegar is a liquid with a strong smell. Baking soda is a solid in the form of a powder.

Carlo did an experiment to see what happens when vinegar and baking soda are mixed. He put solid baking soda in one cup. He put liquid vinegar in another cup.

Carlo put the vinegar cup inside the baking soda cup. He put the two cups on one side of a balance and mass pieces on the other side. He added mass pieces until the system balanced.

Carlo carefully poured the vinegar into the cup with baking soda. The mixture fizzed and bubbled.

What happened? A **chemical reaction**. The vinegar and baking soda reacted. During the reaction, new materials formed. One of the new materials was a gas. The gas that made all the bubbles was a new material. Where did the gas come from?

Baking soda and vinegar

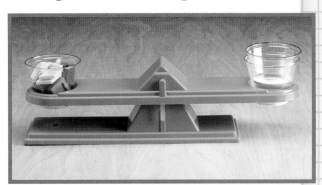

Mass pieces equal to the mass of the baking soda and vinegar

Carbon dioxide gas forms when vinegar and baking soda are mixed.

51

The particles in the vinegar and baking soda combined in new ways during the reaction. One new combination formed the gas **carbon dioxide**. That's where the gas came from. The gas was a new material that formed when vinegar and baking soda reacted.

After the fizzing stopped, Carlo looked in the cup. There was no solid baking soda left. He carefully waved his hand over the cup to bring the smell toward his nose. It no longer smelled like vinegar. The new materials had different properties than the starting materials.

Carlo made one more observation. The mass pieces were still in the cup on the balance. He put his two cups back on the balance. The system did not balance. The reaction cup had less mass than it did before. Why?

Gas is matter. All matter has mass. When the carbon dioxide gas formed, it went into the air. Millions of particles left the cup and went into the air. The material in the cup lost mass.

Particles combine to form new materials. Every different combination of particles makes a different material. The particles rearrange during reactions. New arrangements of particles make new materials.

Carlo made a new material, carbon dioxide gas, by combining baking soda and vinegar.

The mass in the cups after the reaction is less than it was before the reaction.

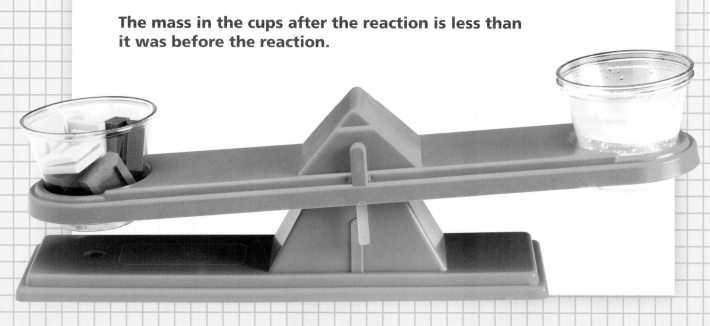

Careers You Can Count On

The metric system is the international system of measurement. The United States does not officially use the metric system. Yet every day, we count on accurate measures. We use measurements at home, work, and play. Here are a few people who use measurements at work.

Scientists

Today scientists around the world use the metric system. In the United States, scientists use it for their research. They use the metric system to measure, collect, compare, describe, and analyze information.

Scientists collect and record information using the metric system.

Pharmacists

Today most medicines are measured using the metric system. Pharmacists measure and label all types of medicines using the metric system. For example, on a bottle of aspirin, the dosage is listed in milligrams (mg).

Pharmacists carefully measure medicines to be sure the dosage is correct.

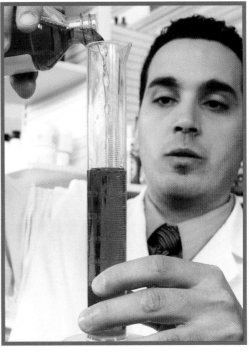

Meteorologists

Meteorologists study Earth's air and weather. They use the metric system to measure temperature and other weather conditions. They also use it to measure the amounts of chemicals in Earth's air.

Meteorologists use the metric system to predict weather conditions.

Biologists

Biologists study living things. They use the metric system to measure and weigh animals and plants. They also use metric measures to map the places where plants and animals live. Biologists who work in zoos use metric measures to help them take care of the animals. Like pharmacists, biologists measure medicines using the metric system.

Biologists measure the growth of plants and animals.

Astronauts

Astronauts are trained to take part in spaceflights. The National Aeronautics and Space Administration (NASA) uses metric weights and measures on all flights.

Astronauts use metric measurements for all their space duties.

Ecologists

Ecologists study the relationships between living things and their environments. Ecologists measure how much pollution is being released into Earth's air and water. They use the metric system for these measurements. Ecologists also use the metric system to measure and map the loss of some environments. This loss is caused by droughts, floods, fires, and natural disasters, as well as by human impacts such as climate change.

Ecologists measure water pollution in metric units.

Archaeologists

Archaeologists study how people lived long ago. They use the metric system when charting and mapping areas they are studying. They also measure and weigh bones and other objects using the metric system.

Archaeologists carefully measure and record each new find.

Chefs and Bakers

Chefs and bakers use many different types of measurements at work. They need to carefully measure all the ingredients in recipes. Bakers also need to know what temperatures their ovens must be. Then they can cook breads, cakes, and other delicious foods.

A skilled chef must measure the right ingredients to prepare a meal.

Engineers

Engineers have to understand many types of metric measurements. They use tape measures, rulers, and other measuring instruments. They also must know how to read blueprints correctly. Blueprints are plans that tell exactly what a building will look like.

Engineers use exact measurements when making blueprints.

Auto Mechanics

Auto mechanics who work on cars from other countries must know the metric system. Instruction books and packaging materials from other countries may use only metric measurements. This also is true of parts used to repair foreign cars and trucks.

Mechanics must use metric tools to repair foreign cars.

Athletes and Sports Officials

Many sports use the metric system for measurement. Track-and-field events, swimming, and skiing are just a few. Runners compete in the 100-, 200-, 400-, and 1,000-meter (m) dashes. Cyclists compete in 10-kilometer (km) races. Divers compete using platforms 10 m high. That's almost as high as a three-story building!

Athletes often compete against one another in different countries. Because most countries use the metric system, it is used at international sporting events as well.

The Boston Marathon is the world's oldest and most well-known marathon. In 1975, the Boston Marathon was the first to include a wheelchair division. In 2012, Josh Cassidy of Canada set a new men's world record at the Boston Marathon with a time of 1 hour, 18 minutes, and 25 seconds.

Amazing Athletic Achievements

These are some world records for international sporting events.

Event	Time/distance	Record holder	Date
MEN'S TRACK & FIELD			
100-meter dash	9.58 seconds	Usain Bolt, Jamaica	August 16, 2009
200-meter dash	19.19 seconds	Usain Bolt, Jamaica	August 20, 2009
Long jump	8.95 meters	Mike Powell, USA	August 30, 1991
WOMEN'S TRACK & FIELD			
100-meter dash	10.49 seconds	Florence Griffith Joyner, USA	July 16, 1988
200-meter dash	21.34 seconds	Florence Griffith Joyner, USA	September 29, 1988
Long jump	7.52 meters	Galina Chistyakova, USSR	June 11, 1988

Veterinarians

Veterinarians use the metric system to measure medicine for animal patients. They also use temperature measurements to see how healthy their patients are.

Veterinarians use the metric system to measure the growth of animals.

Teachers

Science and math teachers help students learn about the metric system. Nearly all the nations of the world use the metric system. That is why it is important for people in the United States to understand and feel comfortable with the metric system.

Teachers explain the metric system to young scientists.

Using the Metric System

Would you believe that we use the metric system already? We use it every single day.

When we talk about electricity, we talk about watts. Watts are metric units.

We buy liters of soft drinks.

We measure medicine and vitamin dosages in milligrams.

Hunt for Metrics!

Can you find five things around your house that have metric measurements on them? Make a list on a sheet of paper.

Science Safety Rules

1. Listen carefully to your teacher's instructions. Follow all directions. Ask questions if you don't know what to do.

2. Tell your teacher if you have any allergies.

3. Never put any materials in your mouth. Do not taste anything unless your teacher tells you to do so.

4. Never smell any unknown material. If your teacher tells you to smell something, wave your hand over the material to bring the smell toward your nose.

5. Do not touch your face, mouth, ears, eyes, or nose while working with chemicals, plants, or animals.

6. Always protect your eyes. Wear safety goggles when necessary. Tell your teacher if you wear contact lenses.

7. Always wash your hands with soap and warm water after handling chemicals, plants, or animals.

8. Never mix any chemicals unless your teacher tells you to do so.

9. Report all spills, accidents, and injuries to your teacher.

10. Treat animals with respect, caution, and consideration.

11. Clean up your work space after each investigation.

12. Act responsibly during all science activities.

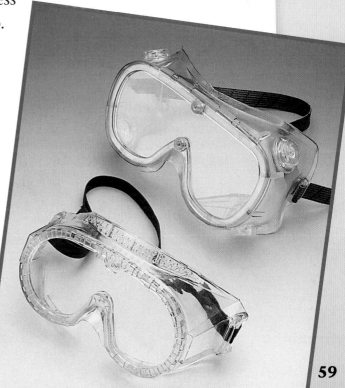

Glossary

attract to pull toward

balanced to be in a stable position

carbon dioxide a gas made of carbon and oxygen

chalk one form of the material calcium carbonate

chemical reaction an interaction between materials that produces one or more new materials that have different properties than the starting materials

cloudy not clear

conserve to stay constant during an interaction. Matter can change, but it's always conserved.

constraint a restriction or limitation

criteria (singular **criterion)** a rule for evaluating or testing something

curved round

data information collected and recorded as a result of observation

direction the path on which something is moving or pointing

dissolve to mix a material uniformly into another

distance how far it is from one place to another

energy the ability to make things happen. Energy can take a number of forms, such as heat and light.

engineer a scientist who designs ways to accomplish a goal or solve a problem

equal the same as

evidence data used to support claims. Evidence is based on observation and scientific data.

experiment a test or trial

float to be supported on the surface of water or to be suspended in air

force a push or a pull

friction a force between objects that are touching each other that opposes their motion, slowing them down

gas a state of matter with no definite shape or volume; usually invisible

gram (g) the basic unit of mass in the metric system

gravity a force that pulls objects toward each other. It is the force of gravity that pulls objects toward Earth's center.

liquid a state of matter with no definite shape but a definite volume

liter (L) the basic unit of liquid volume in the metric system

magnet an object that sticks to iron or steel

magnetic closure something that closes or shuts using a magnet

magnetic field an invisible field around a magnet

magnetic force the force produced by a magnetic field

magnetism a force that attracts iron and steel

mass the amount of material in something

matter anything that has mass and takes up space

meter (m) the basic unit of distance or length in the metric system

mixture two or more materials together

motion the act of moving

natural history the study of plants and animals in nature

observation the act of noticing the properties of an object or event with one or more of the five senses (sight, hearing, touch, smell, and sometimes taste)

pattern a consistent and repeating combination of qualities or behaviors

pole the end or side of a magnet (magnetic pole)

predict to estimate a future event based on data or experience

property something you can observe about an object or a material. Size, color, shape, texture, and smell are properties.

prototype a model

pull when you make things move toward you. Pulling is a force.

push when you make things move away from you. Pushing is a force.

repel to push away from

rotate to turn or spin

rotation the act of turning around as on an axis

rotational force a push or pull given to something turning on an axis

salt a solid white material that dissolves in water; also known as sodium chloride

screen wire mesh used to separate large and small objects

separate to take apart

shaft a long, thin structure that can be used as an axle or axis

sink to go under water as a result of being more dense than water

solid a state of matter that has a definite shape and volume

solution the act of solving a problem. Engineers solve problems. In chemistry, a mixture formed when one or more substances dissolve in another

spin to move by turning around an axis

stable steady

strength the quality of being strong

symmetrical balanced or the same on each side

system two or more objects that work together in a meaningful way

technology practical use of scientific knowledge to solve problems

transparent clear

uneven not level or flat

volume three-dimensional space

wheel-and-axle system a simple machine made of a wheel fixed to a shaft, or axle; both the wheel and axle rotate together

Index